EXPLORING MATERIALS

Materials on the move

Malcolm Dixon

EXPLORING MATERIALS

Editor: Joanna Housley
Designer: John Christopher

First published in 1993 by Wayland (Publishers) Limited,
61 Western Road, Hove, East Sussex, BN3 1JD, England

British Library Cataloguing in Publication Data
Dixon, Malcolm
Materials on the Move. – (Exploring Materials Series)
I. Title II. Series
620.1

ISBN 0 7502 0787 6

Typeset by Strong Silent Type
Printed and bound in Spain by Graficas Estella.

For Joanne and Michelle

Picture acknowledgements
The publishers wish to thank the following for supplying the photographs in this book: Cephas Picture Library 37 (Lance Smith); Chapel Studios 13 (John Henirich); Eye Ubiquitous 12 (Julia Waterlow), 38 (top, Gary Sedcole) (bottom, Bennett Dean); Topham Picture Source 5, 26 (Associated Press), 43 (Associated Press); Wayland Picture Library 22, 23 (Associated Press), 25 (both), 31 (Richard Sharpley), 35, 36, 44; ZEFA 4, 6, 14, 15 (W. Ostgathe), 16, 17 (Big Mike), 24 (James Blank), 30, 32.
Artwork by Peter Bull.

NOTES FOR PARENTS AND TEACHERS

Teachers will find this book useful in implementing the National Curriculum (Science and Technology) at Key Stages 1, 2 and 3. MATERIALS ON THE MOVE provides information and activities which are particularly relevant to Science Attainment Targets 1 (Scientific Investigation) and 3 (Materials and their Properties). MATERIALS ON THE MOVE could be developed in a cross-curricular approach involving science, technology, history, geography, mathematics and English.

Some activities in this book will require the help of a parent or teacher. The section on places to visit will be useful to parents during weekends and school holidays.

Contents

Materials and movement

Today many people travel all over the world using land, sea and air transport. Goods, too, are moved across the world by special forms of transport. Movement through the air, in water and across land has become quick and easy. Yet only one hundred years ago moving people and goods over even short distances was slow and difficult. On land, cars and bicycles are now popular and reliable forms of transport.

ABOVE A jet aircraft prepares for take-off from an aircraft carrier, the USS *Enterprise*.

RIGHT An early motor car from the beginning of this century.

Massive passenger and cargo ships travel the oceans of the world. Jet aircraft carry hundreds of passengers, at great speeds, over vast distances. Humans have even been able to travel through space to the Moon.

One reason for this 'transport revolution' is that designers, scientists and engineers have developed new materials. For hundreds of years water transport depended on the use of wood, animal skins, reeds and iron.

Materials used for early air transport were wood, paper, silk and cotton. The first bicycles and cars used wood and iron. When humans started to make metals such as steel and aluminium, they were soon used in different forms of transport because of their strength and light weight. During the last thirty years more new materials – including metal alloys, plastics and composites – have become available. The special properties of these materials can be suited to the needs of a particular task.

This book will develop your knowledge and understanding of the materials we use to enable us to move through air and space, in water and across land. One chapter discusses the ways in which the materials themselves, whether solid, liquid or gas, are actually moving. Practical activities are included so that you get first-hand experience of somc easily obtained materials.

Particles on the move

Everything in the world is made up of matter. Matter can exist in three states: solid, liquid or gas. A solid, such as a piece of steel or plastic, has a fixed shape. A liquid, such as water or oil, does not have a fixed shape but takes on the shape of the container that holds it. A gas, such as air or hydrogen, is shapeless. Gases spread out to fill all the space available to them.

All matter is made up of tiny particles that move all the time. In solids, liquids and gases the particles are held together with differing strengths. In a solid, the particles are arranged closely together in a neat pattern. They are bonded to each other and can only move gently to and fro.

BELOW
When ice melts it changes from a solid to a liquid.

a: solid

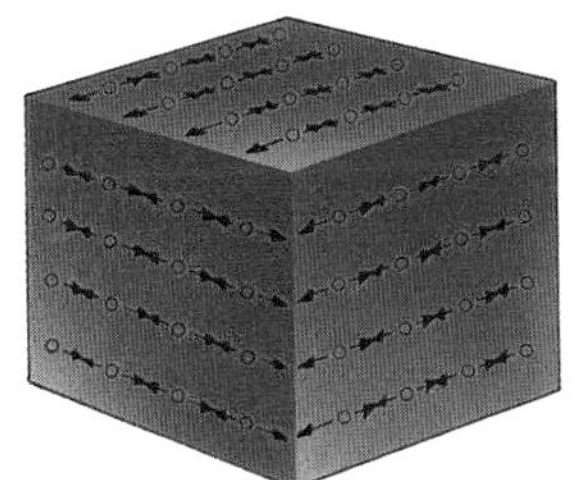

In a solid the particles are close together. They gently vibrate. Solids have fixed shapes.

b: liquid

The particles in a liquid can move in all directions. Liquids take the shape of the container that holds them.

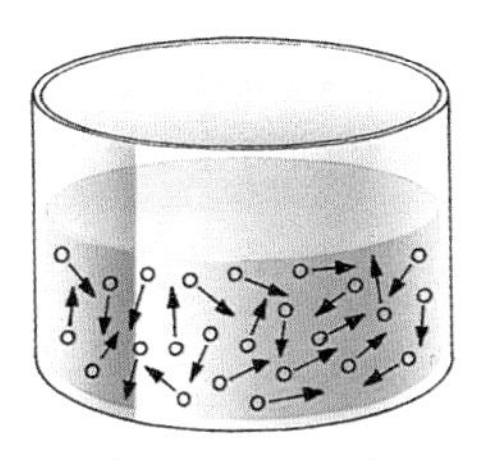

c: gas

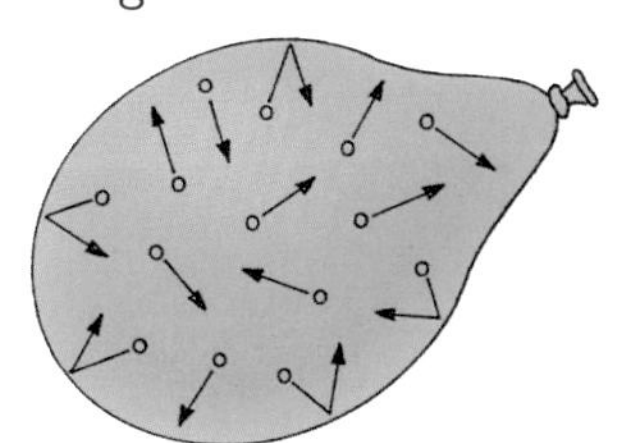

The particles in a gas move at high speed in all directions.

When a solid is heated, the particles are given enough energy to make them move about more. They break away from their neat pattern and the solid changes state to become a liquid. So we can think of the particles within a liquid as being able to move more than the particles in a solid. The bonding between the particles in a liquid is weaker than the bonding of particles in a solid. When a liquid is heated, the particles are given more energy so that they begin to move even faster. Some particles escape from the surface of the liquid and form a gas. The particles within a gas move at high speed in all directions and sometimes collide with each other. These particles are widely spaced with weak bonds between them.

This explanation of how matter behaves is known to scientists as the kinetic theory.

Atoms are tiny particles found in all materials. At the centre of every atom there are two further types of particle, called protons and neutrons, forming an area called the nucleus. Even tinier particles, called electrons, spin around the nucleus.

Usually atoms are linked together in groups called molecules. A molecule of water contains two hydrogen atoms and one oxygen atom. A glass of water could contain about 100 million, million, million, million molecules. These molecules are not strongly attracted to each other.

LEFT Dorothy Hodgkin receiving the Nobel Prize for her outstanding scientific research on materials.

A water molecule. It has the chemical formula H_2O, showing that it has two atoms of hydrogen and one atom of oxygen within the molecule.

A molecule of the gas called carbon dioxide. What is the chemical formula of this gas?

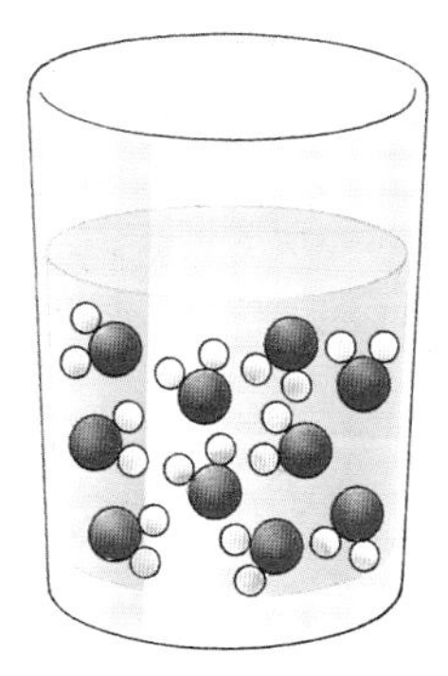

Water molecules are able to move freely in the liquid state.

Once heated, liquid water easily changes state to a gas (steam). Water is said to have a simple molecular structure.

Scientists, such as the Nobel Prize-winning Dorothy Hodgkin, have used X-rays to investigate how atoms are joined together to form materials. Some materials are said to have giant structures. The atoms are bonded together so strongly that the material is difficult to melt or boil. Metals, for example, have giant structures. The atoms of most metals are closely-packed and held together by strong bonds. It is because of their atomic structure that metals are strong, hard, bend easily, conduct electricity and do not melt easily.

Experiment with moving particles

YOU WILL NEED

3 small clear plastic containers
some dried peas

1. Take one plastic container and fill it to the top with peas. Shake the container gently. Notice how the peas are only able to move slightly. The particles in solids are closely packed in this way. What happens when you shake the container harder?

2. Half fill a second plastic container with peas. Shake the container gently. Can the peas move around more than they could in the first container? Which state of matter does this show?

3. Place a few peas in the third container. Shake the container. Watch how the peas move freely in all directions. Shake the container more. Do any of the peas knock into each other? The particles that form gases behave in a similar way.

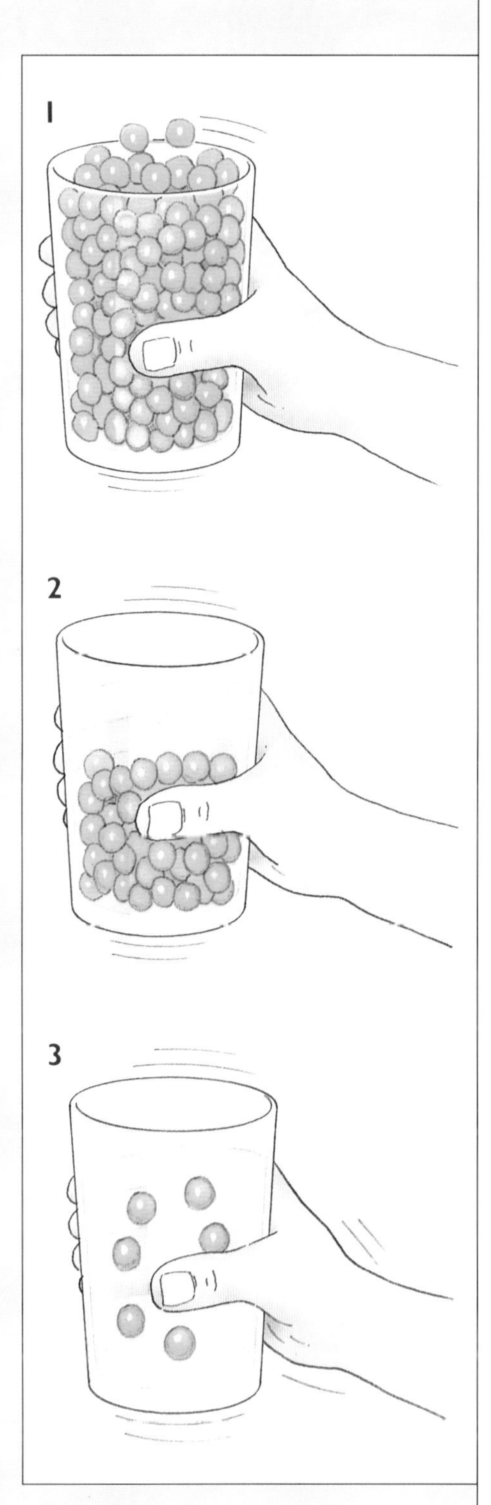

Materials and water transport

The first boats were made using natural materials such as wood and animal skins. For years, boats were, in fact, simply floating logs that people sat or stood on. A more stable craft was made by tying several logs together to form a raft. The ancient Egyptians used papyrus reeds, tied together in bundles, to build rafts. A Norwegian explorer, Thor Heyerdahl, built a raft made from balsa wood to sail across the Pacific Ocean in 1947. This raft, known as the *Kon Tiki,* was similar to rafts used by Polynesian islanders centuries earlier. In 1970, Heyerdahl used a papyrus boat to cross the Atlantic Ocean. His remarkable voyage proved that ancient peoples had been able to travel across the seas to new lands. Rafts, sometimes made by tying bamboo poles together, are still used for fishing in lakes, rivers and seas in many parts of the world.

BELOW The ancient Egyptians built boats like this one, using papyrus reeds.

Inflatable rafts are now used as emergency rescue craft on aeroplanes and ships. They are made from a modern material, rubber, which is strong but elastic. This elasticity allows the material to return to its original shape after it has been bent or stretched. The idea of inflating materials with air was probably first developed by the Assyrians more than 2,500 years ago. They used animal skins blown up with air to make simple floats.

ABOVE These coastguards are using a raft which is inflated with air.

The skins from animals such as buffalo and bison, stretched over a wooden framework, have also been used to construct simple boats in India and North America. The native Americans of North America and the

Aborigines of Australia used bark from birch, elm and eucalyptus trees, fixed to light wooden frameworks, to make canoes. The Inuit people of the Arctic stretched sealskin over a light framework, made from driftwood and bone, to make their fast-moving kayaks.

Modern kayak racing is a popular sport throughout the world. These lightweight craft were at one time built with a wooden framework and waterproof canvas covering. Today they are constructed using a material called glass-fibre reinforced plastic. This light, but strong, material uses glass fibres to give the plastic extra strength. It is an example of a composite material in which two or more materials work together. The latest designs of kayak use materials such as carbon or kevlar to reinforce the plastic and make a lightweight material that is extremely tough. Kevlar is a plastic fibre which is so strong it is also used to make bullet-proof vests.

LEFT Kayaks are built using materials which are lightweight and strong.

By using wood cut into flat pieces called planks, it became possible to build large, deep boats. The Vikings built their longships from wooden planks which overlapped. Other plank-built boats had their planks joined edge to edge to make a smooth surface. The Vikings, Greeks and Romans all used sails and the power of the wind to move their ships.

The great age of wind power came much later, in the nineteenth century. Massive wooden-hulled boats were built and equipped with many sails to carry cargo across the oceans. The sails

BELOW Sailing ships, like this galleon, travelled the oceans over 200 years ago.

used needed to be made of a material which was light in weight, strong, smooth surfaced, and able to resist extreme weather. Most importantly, they had to withstand strong winds! For hundreds of years, sails were made by sewing sections of heavy canvas together. Many ships carried a sailmaker amongst the crew, so that repairs could be made at sea. More recently, canvas has been replaced by manufactured fibres such as Terylene and Dacron. The hulls of modern sailing boats, such as the dinghies and catamarans used for sporting activities, are constructed from plastic reinforced with glass-fibre.

During the 1800s, iron began to replace wood as a material for building the hulls and decks of ships. It was longer lasting than wood, but rusted quickly in sea water. In the late 1800s, steel gradually replaced iron as a shipbuilding material. Huge steel-hulled ships were built by joining steel plates together. Later, luxurious liners were constructed to carry large numbers of passengers across the oceans. Massive supertankers and container ships now carry enormous cargoes from port to port around the world. It was only possible to build such large ships, which have to resist the fierce battering of the sea, because of the great strength of steel and the design of the ships.

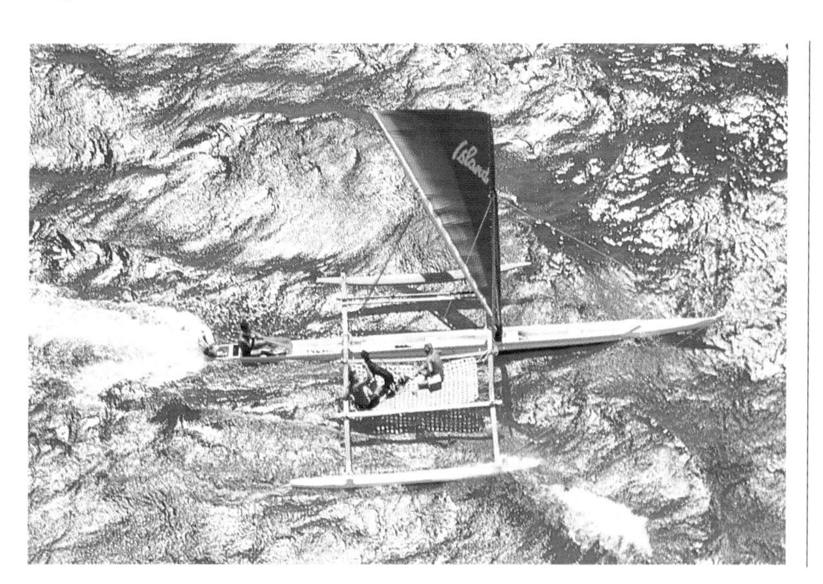

LEFT Modern catamarans are often built using plastic reinforced with glass-fibre.

More specialized ships, such as submarines, aircraft carriers and icebreakers, are also built using steel plates, but are designed with extra-strong hulls. Minesweepers, which are warships that are used to seek out mines, are built with plastic hulls so that they do not attract magnetic mines. Other warships have been built using aluminium. This makes them lighter and so their speed through the seas is faster.

BELOW Submarines have very strong steel hulls.

Experiment with floating and sinking

YOU WILL NEED

a plastic tank or bowl
a spring balance
a plastic jar with a lid
string
marbles
scissors

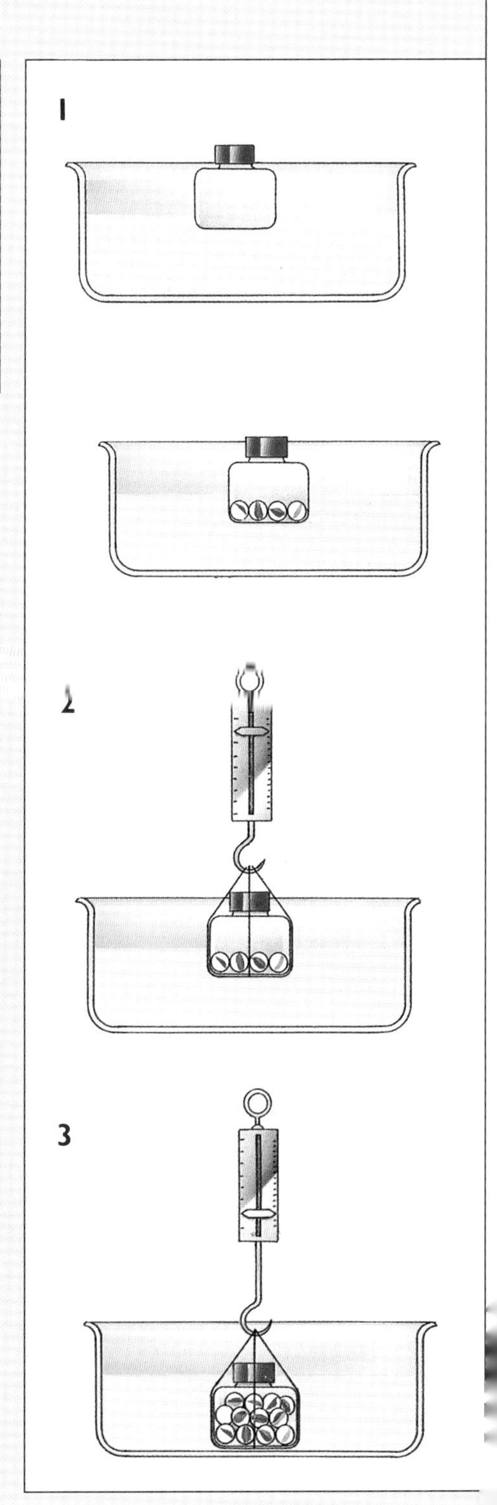

1. Half fill the plastic tank or bowl with water. Does the plastic jar float in water? Place some marbles in the plastic jar and replace the lid. Do you think the jar will float now? Add more marbles. Does the jar sink further into the water?

2. Weigh the jar and marbles from a spring balance. Lower the jar into the tank of water. What is the weight now?

3. Add enough marbles to make the jar sink. The jar is pulled down by gravity. We call this force the weight of the object.
When an object is immersed in water a force called upthrust pushes upwards on the object. The weight of a floating ship equals the upthrust of water on it.

Make a paper raft

YOU WILL NEED

paper straws
(6mm diameter if possible)
paper glue
weighing scales
scissors
a ruler
a bowl of water
Plasticine

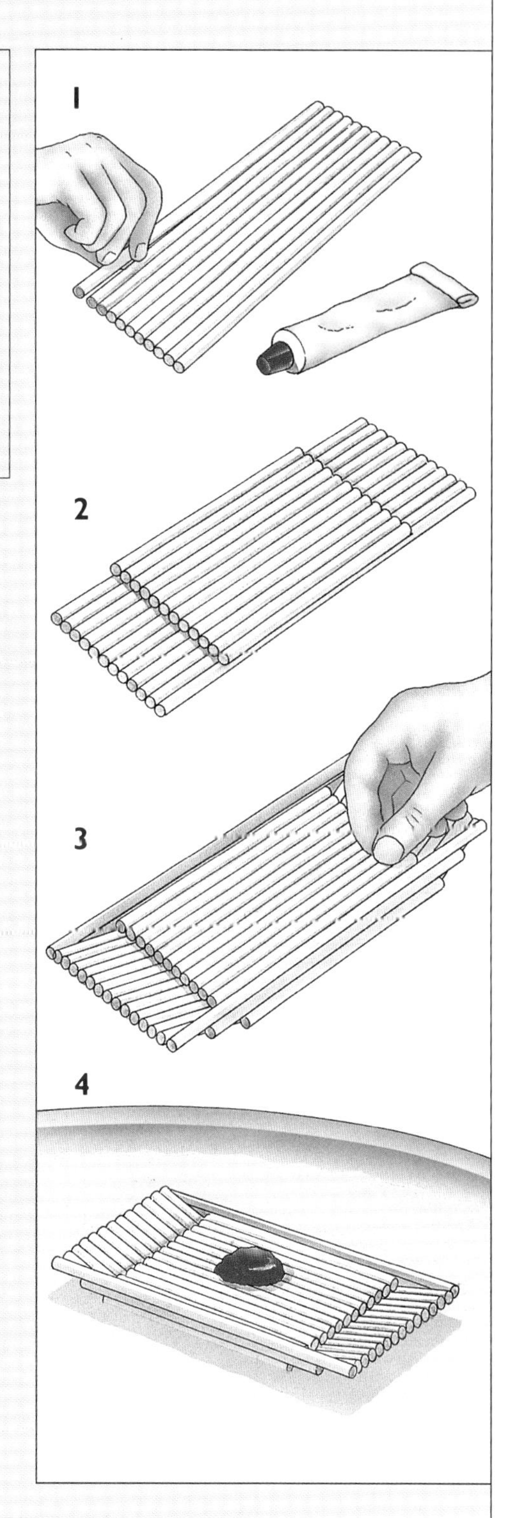

1. Glue twelve straws together to form the base of your raft.

2. Take a further twelve straws and cut them so that they are about 6 cm shorter than the first group of straws. Glue these straws on top of the base layer. Leave them to dry.

3. Fold up the front (bow) and back (stern) straws of your raft. Take two straws and glue them to one side of your raft. Do the same on the other side. Leave them to dry.

4. Use the scales to find the mass of your raft. Place the raft in a bowl of water. Does it float? Place some Plasticine on the raft. Add Plasticine until it sinks. Find the mass of this amount of Plasticine. How much greater is it than the mass of the raft?

Build a model sailing dinghy

YOU WILL NEED

a plastic cooking oil bottle	scissors
a cotton reel	a polythene bag
a hacksaw	a bench hook
a craft knife	sticky tape
strong glue	dowel rod

- **YOU WILL NEED AN ADULT TO HELP YOU**

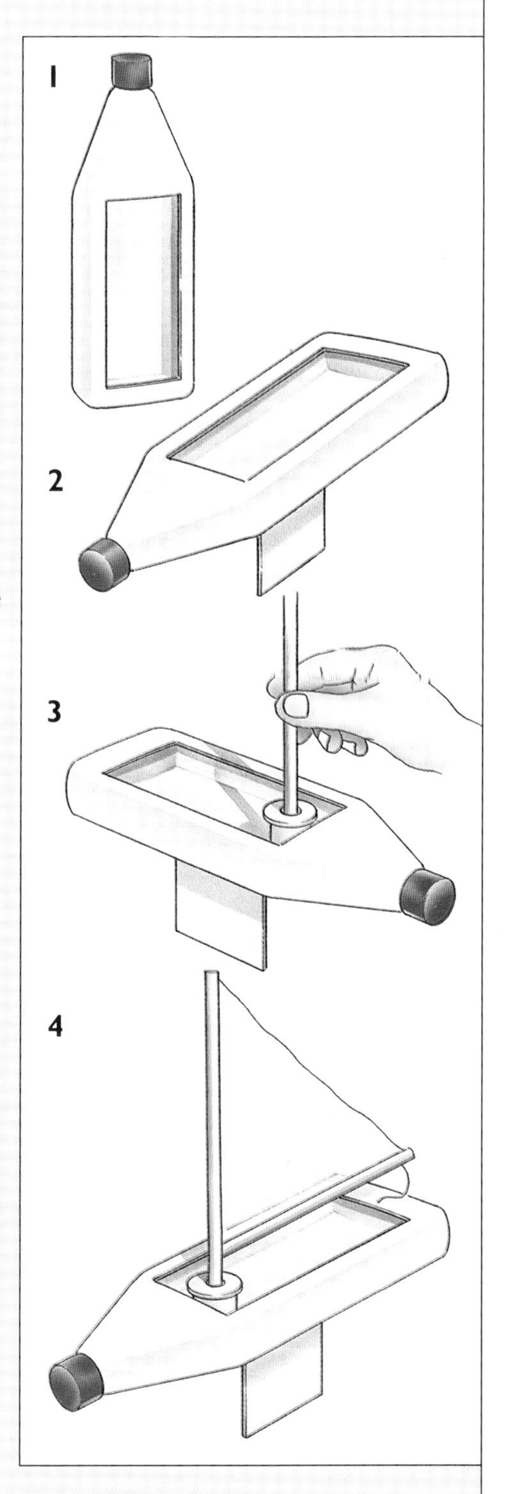

1. Wash the inside of the bottle with soapy water. Allow it to dry and replace the cap. Ask an adult to cut out a section, as shown, from a side of the bottle.

2. Ask an adult to cut a slit in the base of the bottle. Slide the piece of plastic through this slit. It should be a very tight fit. What do you think is the purpose of this daggerboard?

3. Glue a cotton reel inside the bottle as shown. Cut a length of dowel rod and glue it into the cotton reel to form a mast.

4. Make a sail using the polythene bag, as shown. Use dowel rod for the boom (the pole at the bottom of the sail). Use sticky tape to hold the boom in place and to attach the sail to the mast. Fix a length of cotton from the boom to the boat. Test your boat on calm water on a day with a gentle breeze.

Build a motor-driven catamaran

YOU WILL NEED

2 identical washing-up-liquid bottles	a plastic propeller
an electric motor	a hacksaw
a 4.5 volt battery wire	wood glue
1 cm-thick balsa wood	a bench hook
2 paperclips	small pins
elastic bands	a ruler
	scissors

- **YOU WILL NEED AN ADULT TO HELP YOU**

1. Ask an adult to help you cut a piece of 1 cm-thick balsa wood measuring about 25 cm by 10 cm. Use elastic bands to attach this piece of wood to the two empty bottles.

2. Measure the length of the propeller. Cut or build a balsa wood block which is the same height as half of the propeller length. Glue this block to the balsa wood deck. Cut a small length of 1 cm-thick balsa wood. Glue and pin this to the wooden block as shown.

3. Fit the propeller on to the motor spindle. It should be a tight fit. Use an elastic band to hold the motor in place. Connect wires from the motor to a battery. Glue the battery behind the motor mount.

Test your model in a bath of water. Can you make it go forwards and backwards?

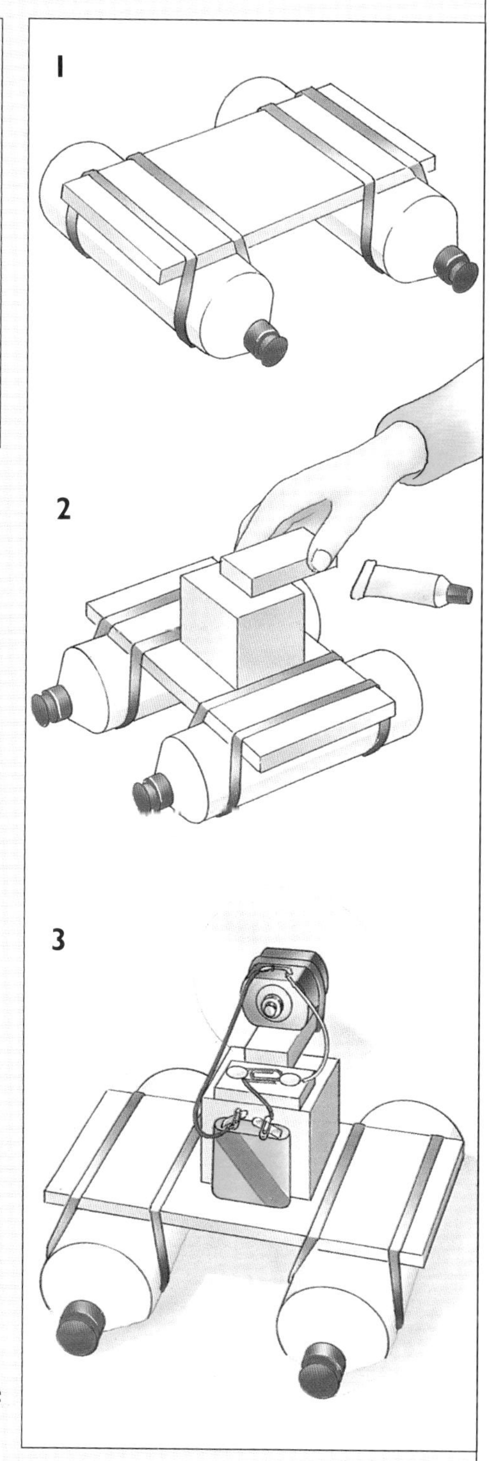

Materials in the air

ABOVE The Montgolfier brothers built balloons using paper, linen and cotton. One of their balloons, filled with hot air, made the first successful human flight.

For thousands of years people have been fascinated by the idea of being able to fly. The first scientific studies of the flight of birds were probably made by the great Italian engineer and artist Leonardo da Vinci in the fifteenth century. Leonardo made many drawings of machines designed to enable humans to fly. These machines were to be built from materials available at that time, such as wood, metals, fabric, reeds and ropes. It is not certain how many of these machines were ever made, but they did give ideas for flying machines that were built later.

In 1783 Joseph and Étienne Montgolfier, the sons of a French paper maker, constructed a balloon using sections of linen strengthened by several layers of paper, ropes, wood and wire. Their balloon was inflated by hot air made by burning wood and straw beneath it. Since hot air is lighter than cold air, the balloon quickly rose and stayed in the air for ten minutes.

A few months later, the Montgolfier brothers built and launched a balloon made from cotton but coated with a substance called alum to try to make it fireproof. This balloon, which was 25 m high and 15 m in diameter, had a circular gallery attached to its base and held up by rope. A burner, to provide the hot air, was suspended from the inside of the balloon using chains. On 21 November 1783, this balloon, carrying two men in the gallery, rose over 1,000 m and travelled about 10 km. By using a lighter-than-air craft, the first successful human flight had been made.

In 1852 a French engineer, Henri Gifford, constructed the first 'dirigible' or steerable balloon. This whale-shaped balloon was filled with coal-gas and carried underneath it a small platform on which there was a steam engine. The steam engine and a rudder were used to power and steer the balloon. The first metal dirigible was built by an Austrian engineer in 1897. It had an aluminium framework entirely covered in extremely thin sheets of aluminium. Dirigibles became known as airships, and they were used in the First World War (1914-18). In the 1920s huge airships, some over 200 m long, carried passengers across the Atlantic Ocean. But in 1937 a massive airship called the *Hindenburg*, inflated with flammable hydrogen, exploded. Thirty-five people were killed. This ended the development of this form of air transport.

ABOVE The tragic accident involving the German airship *Hindenburg*.

LEFT Modern hot-air balloons are often made from nylon coated with polyurethane.

RIGHT Orville and Wilbur Wright flying their glider at Kitty Hawk in 1902.

OPPOSITE BELOW Concorde shows you how much aeroplanes have changed since the Wright brothers' first flight.

However, within the last fifteen years new versions of the airship have been built. Modern materials, such as plastic composites and carbon-fibre, have been used. The airships are filled with a gas called helium. This is lighter than air, does not catch fire easily, and is therefore much safer than hydrogen. Hot-air ballooning is now a popular sport. Hot-air balloons, some with spectacular shapes, are also used for advertising products. Propane gas burners provide the heat. These modern balloons are often made from nylon coated with polyurethane.

Attempts to build heavier-than-air flying machines were pioneered by Sir George Cayley early in the 19th century. Cayley carried out many experiments and in 1853 built a glider that carried one of his servants a short distance in the air. By 1891 a German inventor, Otto Lilienthal, had built a glider in which he was able to fly 250 m. Lilienthal's glider designs, which were similar to modern hang-gliders, allowed him to make over 2,000 flights. In 1848, an Englishman, John Stringfellow, showed that powered flight was possible using a model aeroplane and a lightweight steam engine.

But it was not until 1903 that Orville and Wilbur Wright made the first controlled powered flight over a distance of about 36 m. For this flight, at Kitty Hawk in the USA, the Wright brothers used a small petrol engine, weighing around 82 kg, made largely from aluminium. Six years later, Louis Blériot made the first powered flight across the English Channel.

Rapidly, aeroplanes were developed for military use in the First World War, and later for passenger flights. These early aircraft were built with wooden fuselages (aeroplane bodies) and wings. Timbers such as willow, spruce, ash and hickory were lightweight, strong and easy to work with. Sometimes wire was used to give additional strength to the structure. The wooden framework of the aircraft was then covered in fabric such as silk, cotton or linen. Later, thin plywood began to be used as a covering material.

Modern passenger aircraft now depend upon aluminium and its alloys for over 70 per cent of their mass. The thin sheets of metal which cover the fuselage and wings are sometimes made from clad alloys. In this process, an alloy sheet is covered on both sides with a thin layer of pure aluminium. This 'three-ply' metal makes a long-lasting outer covering for an aircraft.

Concorde, which can fly at twice the speed of sound (2,170 kph), is largely made from an aluminium alloy called RR.58. This special alloy is able to withstand the terrific forces on it while travelling at such high speeds. Military jet aircraft contain a large amount of titanium and its alloys. This metal is very strong, resists corrosion and is extremely light. Some military aircraft which are capable of flying at three times the speed of sound, such as the Lockheed Blackbird, contain over 80 per cent of their mass in titanium alloys.

Aircraft technology is always improving. Ceramics, new alloys and composite materials will be used more in aircraft construction. Military aircraft, such as the Stealth bombers, have already demonstrated the value of composite materials. These advanced aircraft are able to travel at extremely high speeds and are almost invisible to enemy radar. Gliders are now made from the strong and light composite, glass-reinforced plastic. Modern jet passenger aircraft include carbon-fibre composite materials within their structure.

BELOW Composite materials are used in this F-117A Stealth fighter plane.

Build a hot-air balloon

YOU WILL NEED

4 large sheets of tissue paper	a ruler
PVA glue	newspaper
a felt-tip pen	a hair dryer
	scissors

- **YOU WILL NEED AN ADULT TO HELP YOU**

1. Take a sheet of newspaper and mark out a template (outline) as shown.

2. Place the four sheets of tissue in a pile. Place the newspaper template on to the tissue pile. Cut around the edge of the newspaper through all the sheets of tissue paper.

3. Take two sheets of tissue paper. Glue them together along one edge. Repeat this with the other sheets of tissue paper until all four sheets are glued edge to edge. If there is a small opening at the top of the balloon, glue a small piece of tissue paper over it.

4. Ask an adult to inflate the balloon with hot air from a hair dryer. Hold the hair dryer near the neck of the balloon. Watch how the balloon rises.

Hot-air balloons need to be made of light materials. Hot air is lighter than cold air. When the balloon is filled with hot air, it rises above the cold air surrounding it.

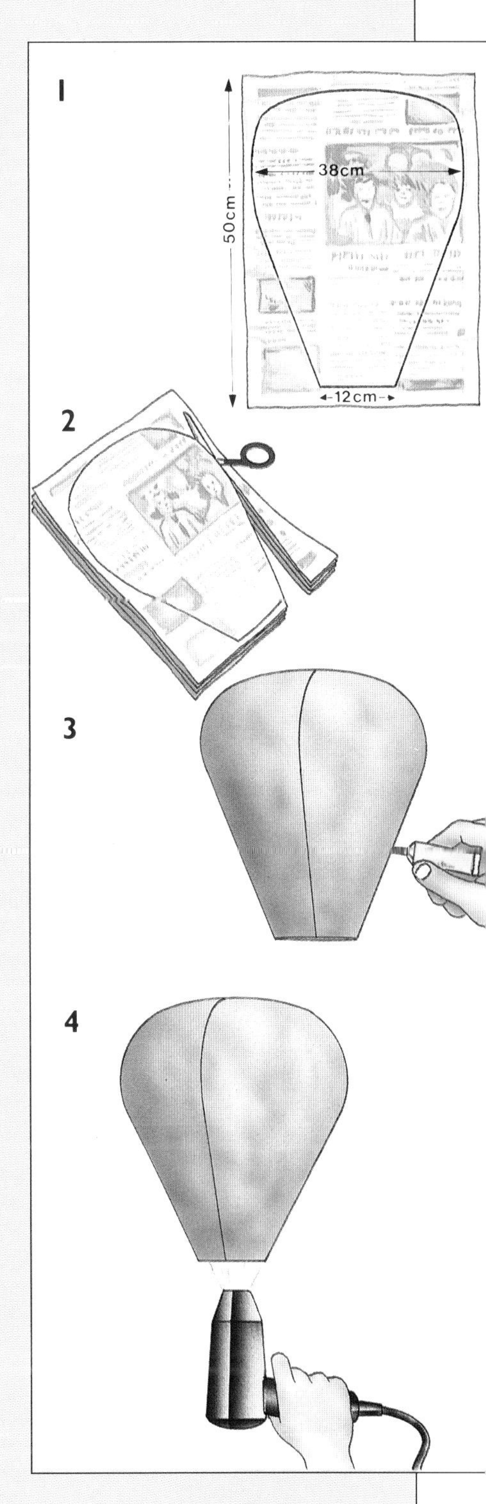

Build a wooden glider

YOU WILL NEED

balsa wood (about 2 mm thick)
a craft knife
glue
a paperclip
a wooden board
thin card
glass paper
scissors

- **YOU WILL NEED AN ADULT TO HELP YOU**

1. Copy the outline of the fuselage and tailfin, shown right, on to a sheet of balsa wood. Ask an adult to cut out the shape carefully with a craft knife. Do the cutting on a wooden board. Cut an 8 cm slit in the fuselage for the front wings. Cut a 5 cm slit for the tailplane.

2. Copy the front wing, shown right, on to balsa wood and carefully cut it out.

3. Copy the tailplane on to balsa wood and carefully cut it out.

4. Assemble your glider by putting the front wings through the slit in the fuselage. Glue it into position. Place the tailplane through the slit below the tailfin and glue it in position. Let the glue dry. Use glass paper to make smooth edges on the wings, tailfin, fuselage and tailplane.

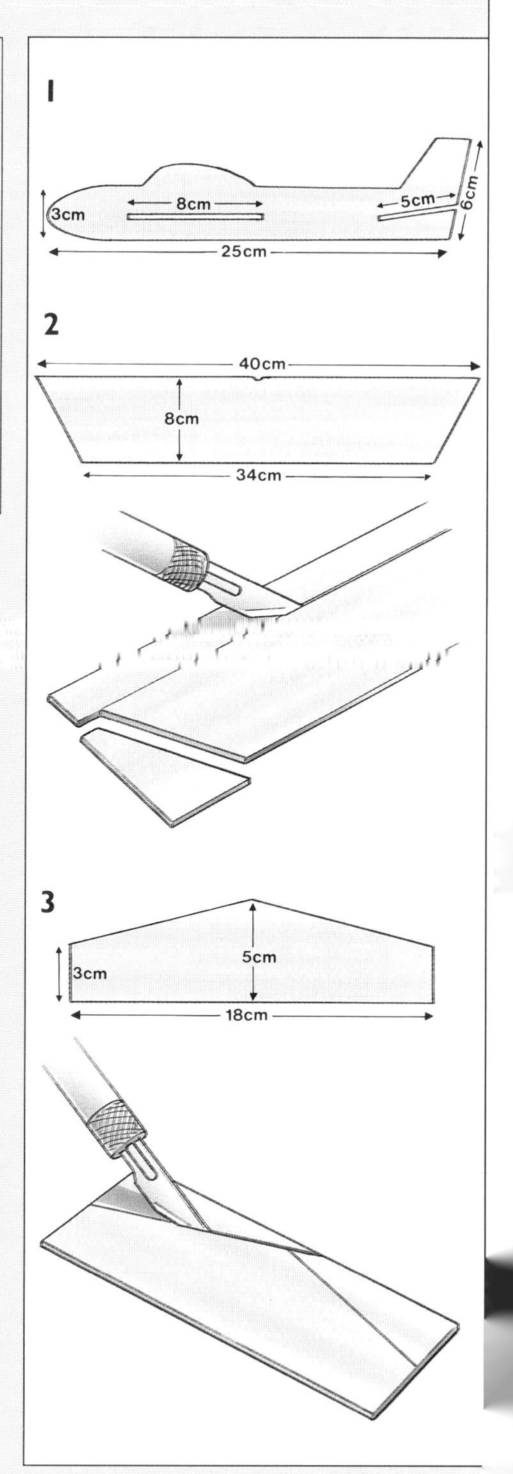

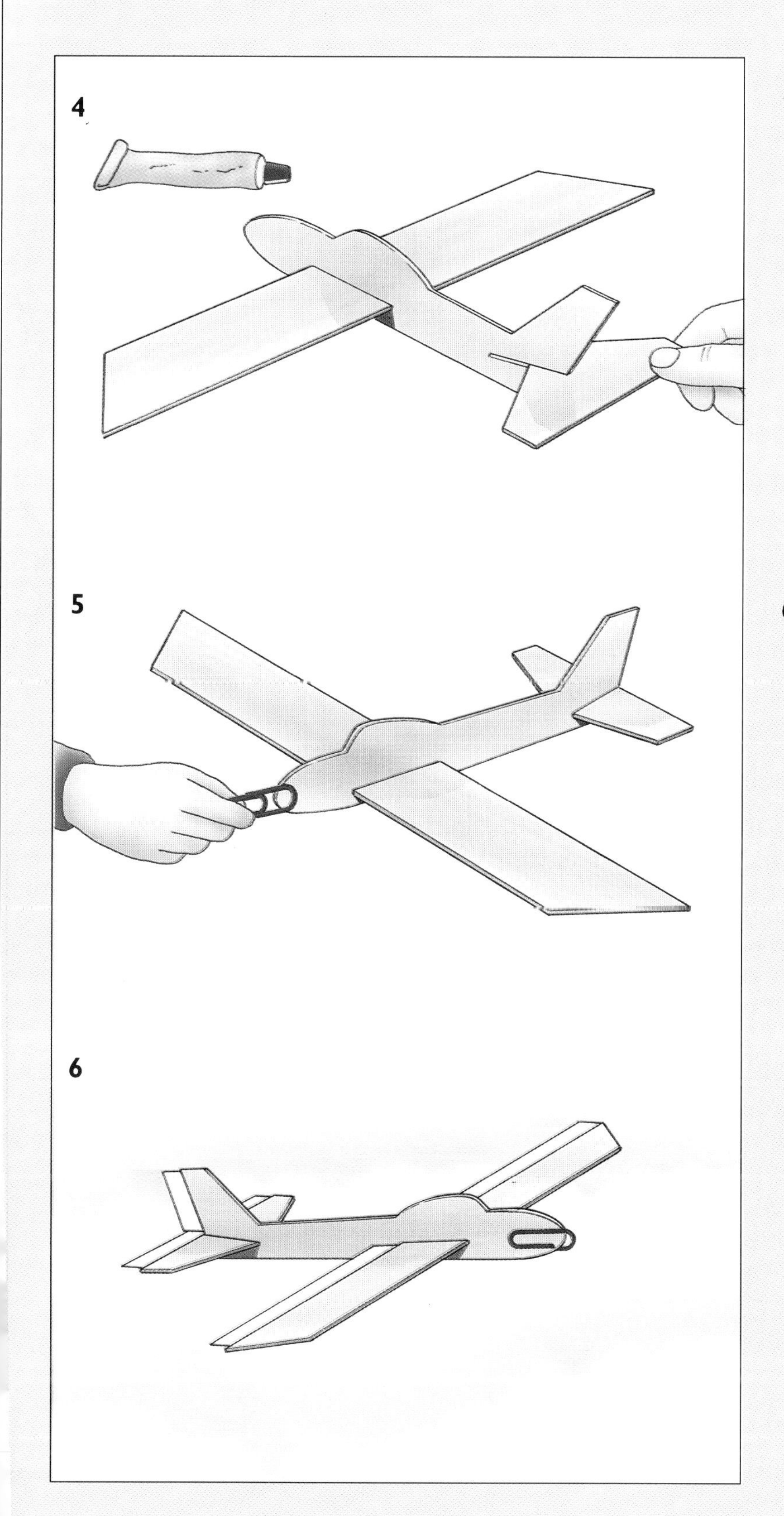

5. Fix a paperclip to the front of your glider. Hold the fuselage beneath the wings and throw the glider forward. Move the paperclip, backwards or forwards, until your glider makes a level flight through the air.

6. Use thin card to make ailerons, elevators and a rudder for your glider. Glue them in place, as shown, and allow the glue to dry. These flaps allow the pilot to control the direction an aircraft takes through the air. Move the rudder to the right. How does your glider fly? What happens when you move the elevators upwards? Set one aileron up and the other down. Launch the glider. What happens?

Materials and land transport

The first bicycle was a wooden-framed machine built in the 1790s. To ride the machine a person sat on the frame and pushed with their feet on the ground. A Scottish blacksmith, Kirkpatrick Macmillan, built the first pedal-pushed bicycle in 1839, using iron as well as wood for the frame and wheels. Since then spoked wheels, air-filled tyres, and gears have been developed to make the bicycle faster and more comfortable.

BELOW Some early bicycles had a large front wheel and a much smaller back wheel. This type of bicycle was nicknamed a 'penny-farthing'.

Today, there are more than 800 million bicycles in use in the world. Bicycle frames can now be made from metals such as steel, aluminium and titanium, or from composite materials. Which materials are used will depend upon the purpose of the bicycle and the cost. For everyday use, most bicycles throughout the world are still made from steel tubing. The frames of these bicycles are strong and long-lasting. Some children's bicycles have aluminium frames, making them lightweight. A bicycle with an aluminium frame will not last as long as a steel-tubed bicycle.

ABOVE A bicycle traffic jam in Beijing, China.

Mountain bikes, invented in the 1970s, need to be light but strong so that they can be used over any kind of ground. Some of the lightest mountain bikes have frames made from carbon-fibre and aluminium alloys. The wheels are made from aluminium alloy with stainless steel spokes. The tyres are reinforced with kevlar to make them stronger.

Racing cycles are built for speed. They need to be very light, strong and streamlined. Metal alloys have been used to build the frames for these bicycles. In the 1990s, carbon-fibre has been shown to be superior to other materials for frame building. It is far stronger, lighter and stiffer than aluminium or steel. Chris Boardman used a 'superbike', with a frame moulded from carbon-fibre, to win a gold medal for cycling in the 1992 Olympic Games. The pedals, seat and chainset were made from titanium.

BELOW Mountain bikes are used on all types of ground.

Investigate bicycle materials

YOU WILL NEED

a bicycle
black paper
a small magnet
chalk or talcum powder
a notebook and pencil

1. Make a list of the parts of the bicycle which are made from metal. Can you identify which metal is used? Are there any labels to help? Use a magnet to identify those parts made from iron or steel.

2. List the parts made from rubber.

3. Can you find anything made from plastic or leather?

4. Are there any parts of the bicycle which are rusting? Do you know why this happens? How can it be prevented? Do all metals rust?

5. Rub chalk or talcum powder on to part of one tyre. Move the chalked part over some black paper. Look at the tyre pattern. Why do you think it is like this? Do other cycle tyres have similar patterns? Do the same with a friend's bicycle.

LEFT What materials were used to make your bicycle?

Investigate air in tubes

YOU WILL NEED

a bicycle inner tube and valve
a bicycle pump
a long balloon
a balloon pump

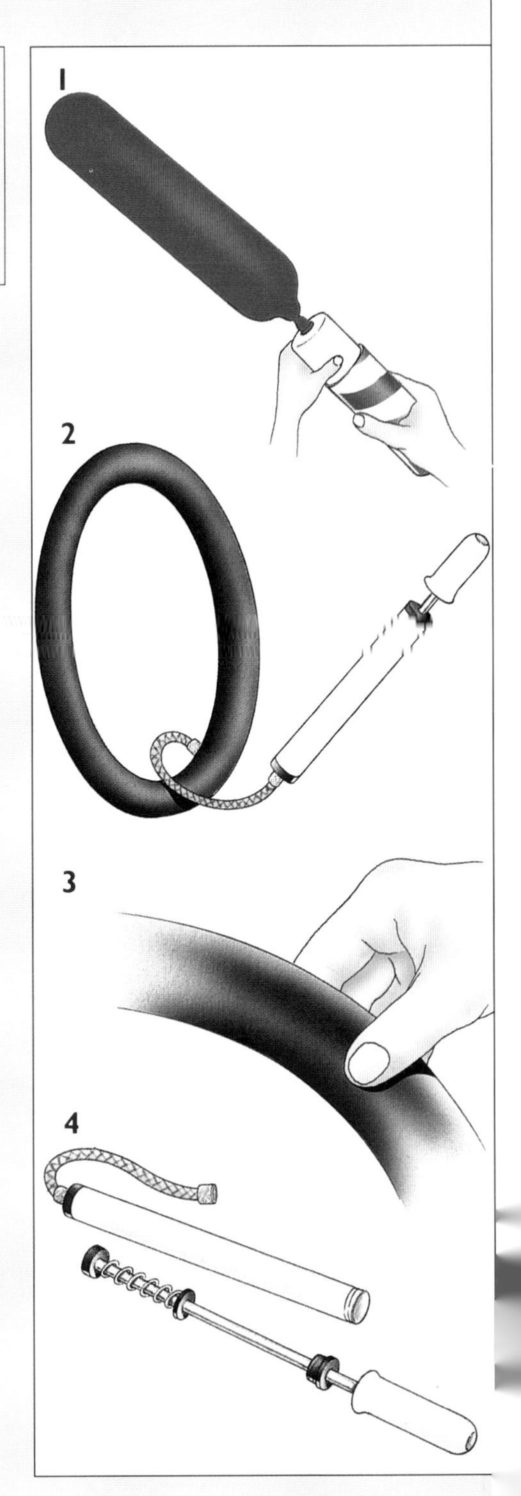

1. Use the balloon pump to inflate a long balloon. What happens when you remove the pump?

2. Attach the bicycle pump to the valve of the inner tube. Pump air into the inner tube. How many 'pumps' does it need to inflate the tube? Remove the bicycle pump. What happens to the air in the inner tube? Can you suggest how the valve works?

3. Squeeze the inner tube with your hand. What happens when you release your grip? Why are air-filled tyres better than solid ones?

4. Take the bicycle pump to pieces. Can you explain how it works? (Remember to put it back together afterwards.)

Watch as an adult uses an air pump, at a garage, to inflate car tyres.

ABOVE Karl Benz and his wife in a car built in 1894. Notice the wheels have solid rubber tyres.

The first motor cars were simply horse carts with engines. The bodywork of these 'horseless carriages' was made from wood. Iron or wood was used to make lightweight wheels and these were covered with solid rubber tyres. Riding in these cars was later made more comfortable by using iron springs, similar to those used in horse-drawn carriages, and air-filled tyres. In Germany in 1885, Karl Benz designed and built the first car to be sold to the public. This car had three wheels with a small engine at the rear.

Soon cars were developed in Britain, France and the USA based on a chassis, to which the bodywork and other parts of the car were attached. The chassis was made from two pieces of steel running the length of the car. More steel was fixed across these lengths to make a strong framework.

ABOVE Computer-controlled robots are used on car production lines.

At first the bodywork was made from wood, but this was changed to steel panels which were welded together. Some expensive cars, such as the Rolls-Royce, had their bodywork made from more expensive but lighter weight aluminium sheeting.

In Europe, the early cars were built by skilled engineers, but in the USA, Henry Ford devised a method of making his Model T Fords using a moving production line. With his mass production methods he built cars quickly and sold them cheaply. Cars are still mass-produced from thousands of different parts, but production lines now use computer-controlled robots instead of human workers for many tasks.

During the 1970s, car manufacturers were forced to re-think the methods and materials used to produce cars. It was predicted that energy costs for the running and production of vehicles would soar. In the USA it had been fashionable for cars to be very long, heavy, chrome plated, and to use vast amounts of fuel. There was now a need for cars which were smaller, lighter, and used less fuel. It was also important for cars to be safe, long-lasting and cheaper.

New forms of steel have been developed and are being used in the cars of the 1990s. More than 50 per cent of the steels now used in cars were unavailable six years ago. These new steels are stronger and can be shaped more easily than those used before. They can be coated with special materials to protect them from corrosion for up to ten years. Cars are built with strong, steel safety cells which surround the driver and passengers. To the front and rear of this safety cell there are 'crumple zones'. In an accident the crumple zones collapse, absorbing the force of the crash, while the safety cell protects the passengers.

Aluminium alloys can be as strong and stiff as steel, but half the weight. They are used in car engines, in radiators, for wheels and other parts. Aluminium could be used more in bodywork. However, compared to steel, aluminium is more expensive, less hard and difficult to weld. Magnesium alloys, which were used in the famous Volkswagen 'Beetle' cars, have been suggested as an alternative to aluminium alloys.

More use is being made of tough plastics and plastic composites for parts of cars such as bumpers. Some car-makers use glass-reinforced plastic for body shells. This material is rust-resistant and easily repaired but does not have the great strength and stiffness of steel. Many modern racing cars use carbon-fibre composite materials for their one-piece body shells.

LEFT This is an extremely long American car, which uses up huge amounts of fuel.

Composites using steel are being developed and will be used in the cars of tomorrow. Wood, in the form of a special plywood, has recently been used in a car developed for use in Africa. This 'Africar', built to cope with poor roads, has both its chassis and bodywork made from the plywood.

ABOVE New, lightweight materials are used to build racing cars.

Modern car tyres still look as if they are made from one black material. In fact, they are made from a thick outer layer of hard rubber supported by inner layers made from steel and nylon fibres. Most modern car tyres are said to be tubeless since, once inflated, the air is held in by the inner lining of the tyre rather than a separate rubber inner tube.

At present, millions of cars are scrapped every year. There is great pressure on car manufacturers to find ways to recycle the materials used in cars and to avoid wasting materials while cars are being produced. Some recently produced cars have plastic parts which, it is claimed, are completely recyclable. These parts, such as the bumpers and fuel tanks, are specially marked so that the type of plastic can be identified in the future, for recycling.

LEFT Millions of cars are scrapped every year.

Build an electric powered car

YOU WILL NEED

some lengths of 1 cm-square wood
a bench hook
a hacksaw
a glue gun
PVA glue
a small electric motor
an elastic band
a 6 mm hole punch
a dowel rod, 6 mm in diameter
plastic containers
thick card
scissors
a ruler and a pencil
a cotton reel
paperclips
wire
4 washing-up-liquid bottles

- **YOU WILL NEED AN ADULT TO HELP YOU**

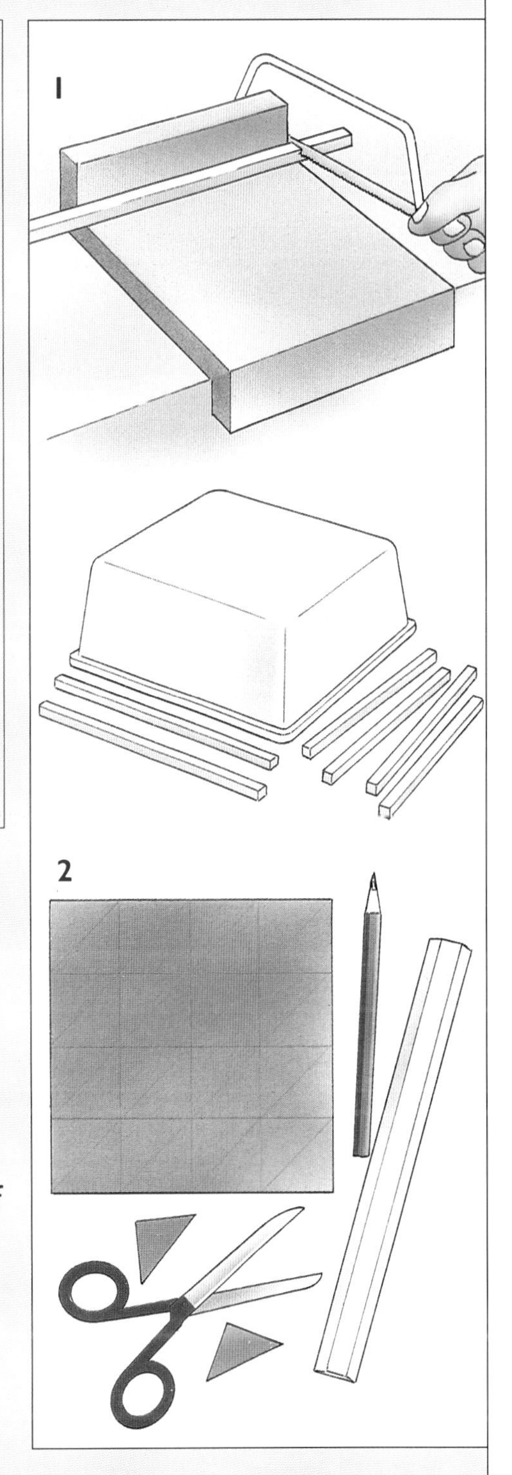

1. Take a large clean plastic container such as an ice-cream container. Measure the length of this container. With an adult's help, use a hacksaw and bench hook to cut two sections of 1 cm-square wood to the same length as the plastic container. Measure the width of the container and, using the same method, cut four lengths of 1 cm square wood to this measurement.

2. Draw horizontal and vertical lines 3 cm apart on a sheet of thin card. Draw diagonal lines across these. Cut out the card triangles.

3. Fix together the six pieces of wood you have cut to make a rectangular chassis as shown. Fix card triangles, using the PVA glue, at each joint. Let the glue dry. Turn the wooden chassis over. Glue card triangles to the other side and leave them to dry.

4. Use thick card to make four axle holders. Cut four rectangles each measuring 6 cm by 4 cm. Punch a 6 mm hole in each card. Glue one card to each corner of the wooden rectangle as shown. Leave the glue to dry.

5. Cut the top and bottom sections from a clean washing-up-liquid bottle as shown. Fit the top into the bottom section to make a small wheel. Make three more wheels in the same way.

6. Use the hacksaw and bench hook to cut two dowel rod axles. Slide the dowel rods through the holes in the axle holders. Make sure the dowel rods can turn easily. You may need to make the holes slightly larger. Fit the wheels on to the ends of the dowel rod axles.

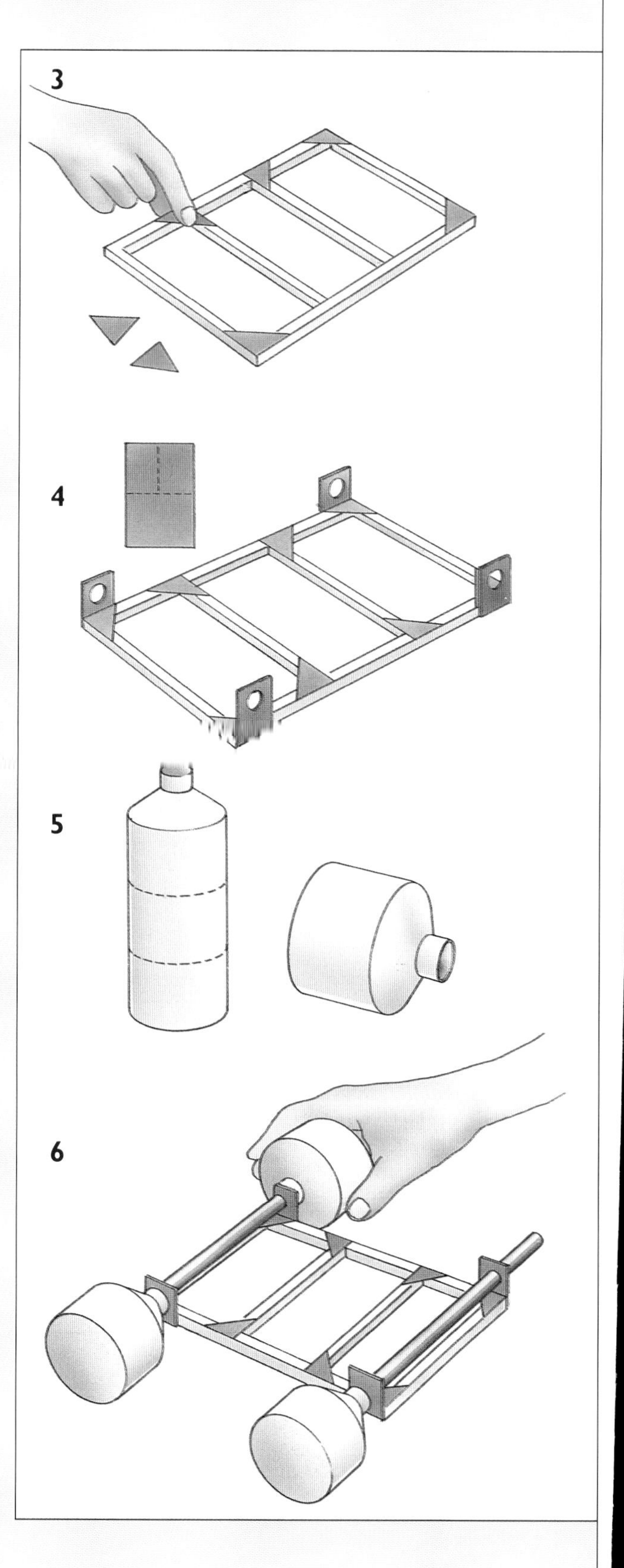

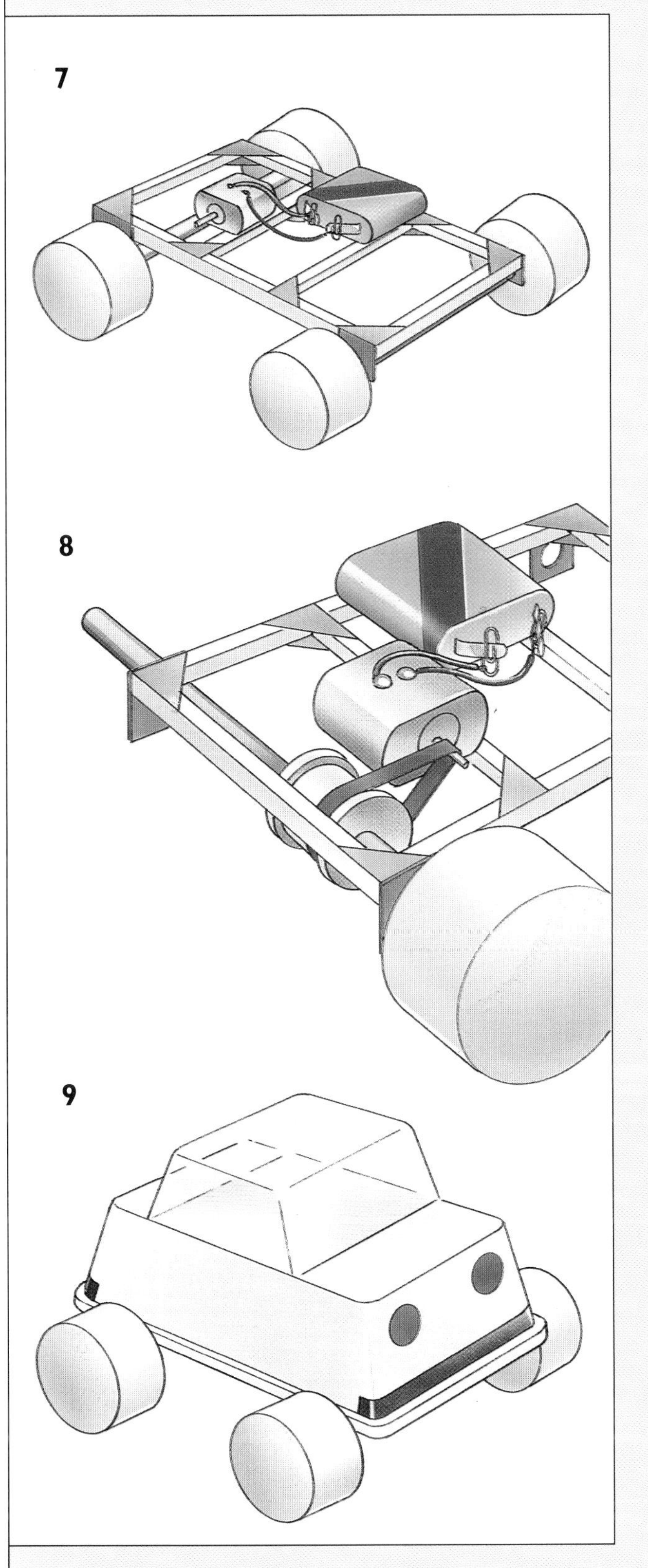

7. Use the glue gun to glue a battery and small motor in position as shown. Use the glue gun to fit a small amount of glue on the end of the motor spindle. Connect the battery to the motor using paperclips and wire.

8. Remove one wheel and slide a cotton reel on to the rear axle. Place an elastic band around the cotton reel. Glue the cotton reel to the axle. Replace the wheel. Position the loose end of the elastic band around the motor spindle. It must not be loose or too tight. Switch on the motor. It should turn the elastic band, the cotton reel and the rear axle. Switch it off. Glue all the wheels to the axles.

9. Glue the large plastic container on to the wooden chassis. Glue a smaller container – a see-through one would be suitable – on top of the larger plastic container. Try to make the body shape more realistic by using card, plastic and kitchen foil for door frames and bumpers. Test your vehicle. How far does it travel on a smooth surface?

Investigate car materials

YOU WILL NEED

a car!
paper
a small magnet
a pencil

1. Look at the picture of a car. Try to identify, on a real car, the materials used for the different parts . Use a small magnet to help you identify those parts made from iron or steel.

2. Visit car showrooms and collect some booklets about different makes of cars. Write to some car manufacturers for further information. Make an attractive display, including photographs and diagrams, and write about the materials used to build cars.

3. Draw a car to your own design. Label the different parts and show what materials you would use for each part.

Look at someone's car to find out what materials it is made from.

Materials in space

During the last forty years humans have been able to build rockets, satellites and other spacecraft to explore space. Powerful rockets have been built which can escape the pull of the earth's gravity. In 1961 a Russian, Major Yuri Gagarin, became the first man to travel in space. Many communication satellites have now been launched into orbit around our planet. They send information, including television pictures and telephone calls, from one side of the world to the other. Since 1969, the American *Apollo* space flights have enabled a number of astronauts to land on the Moon and return to Earth. The re-usable US space shuttle, which is both an aeroplane and a space-craft, is an exciting development.

All of these space vehicles have required the invention of materials able to withstand conditions very different from our own atmosphere. Aluminium and titanium alloys, because of their high strength and light weight, are to be found in many spacecraft. Stainless steel, which is strong and resists heat and corrosion, is also used. Gold is used to protect electronic circuits. Aluminium-lithium alloys, which are light but expensive, and composite materials are used in satellites. In the future, scientists plan to produce new materials in space. We should also remember that space research has produced many new materials which we can use on earth.

RIGHT
The latest materials are used in spacecraft, such as this US space shuttle.

Transporting materials

One hundred and fifty years ago the products that people needed were made or grown locally. Today goods, in the form of solids, liquids and gases, are transported by land, sea and air to all parts of the world.

For example, when oil is extracted from the ground it is called crude oil. It then needs to be transported, often in ships called tankers, to different parts of the world where it is refined. As a result of this process, useful fuels such as petrol and diesel oil are produced. Most plastics are made from crude oil. Unfortunately, moving the oil in tankers across the oceans of the world can be dangerous. Accidents that cause oil spillage damage the environment. Tankers have large steel tanks to hold the oil within their long, box-like hulls. Sometimes the seas are polluted when the tanks have been emptied and are being washed out with sea water.

In March 1989 a giant supertanker called the *Exxon Valdez* went aground in Alaska. From her cargo of 53 million gallons of crude oil, 10 million gallons polluted the beautiful coastline. Millions of fish and thousands of other creatures were killed. In January 1993 another oil tanker, the *Braer*, went aground in hurricane force winds in the Shetland Islands. Thousands of gallons of oil were released, causing an environmental disaster. One possible solution to this problem of moving dangerous liquids safely is to build ships with double-layer hulls. But even this may not be strong enough to stop a ship breaking up when battered by very powerful seas.

BELOW Crude oil is drilled at rigs like this one, and then transported all over the world.

Topic web

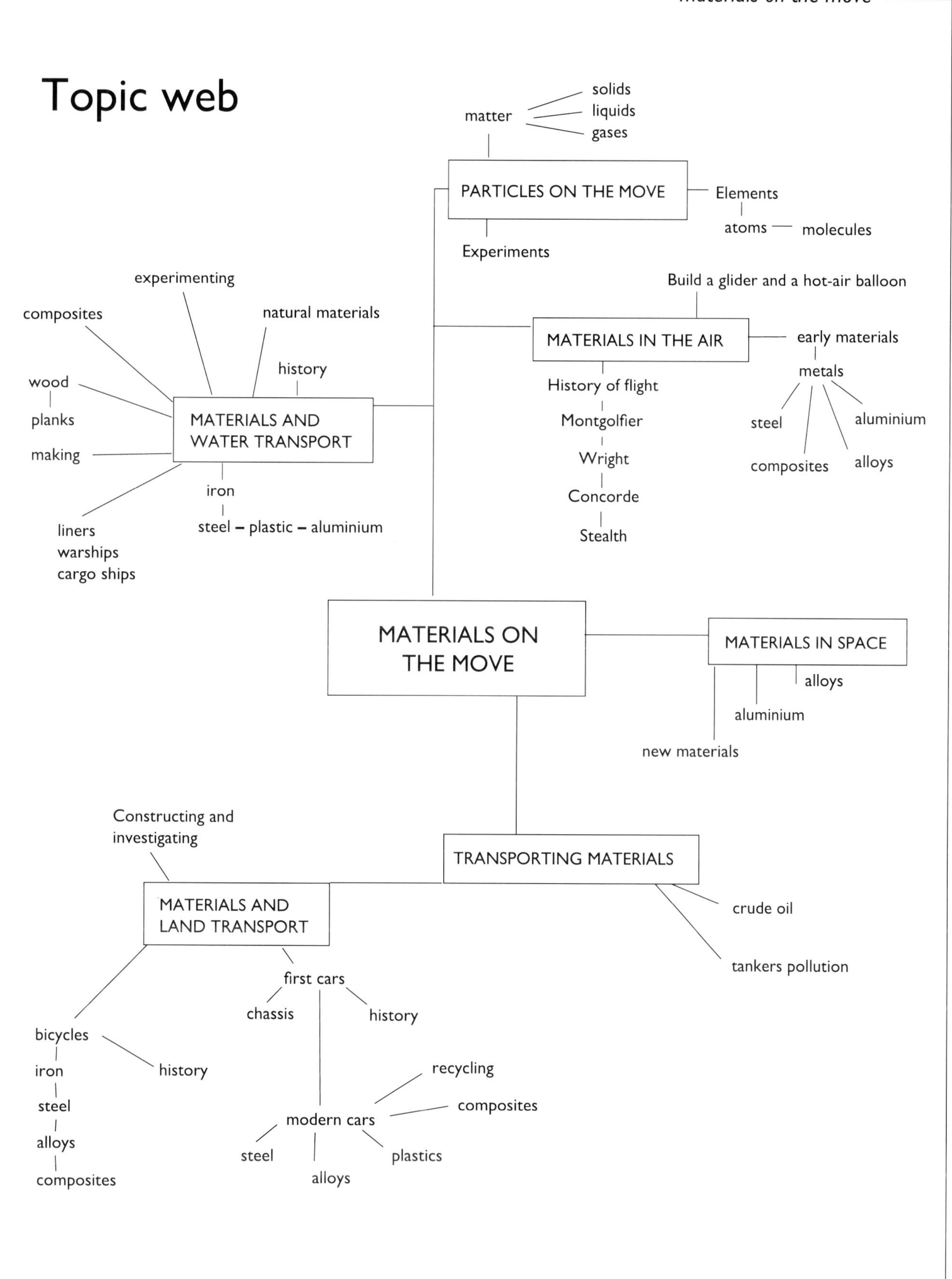
matter
solids
liquids
gases
PARTICLES ON THE MOVE
Elements
atoms
molecules
Experiments
Build a glider and a hot-air balloon
MATERIALS IN THE AIR
early materials
metals
steel
aluminium
composites
alloys
History of flight
Montgolfier
Wright
Concorde
Stealth
experimenting
composites
natural materials
history
wood
planks
making
MATERIALS AND WATER TRANSPORT
iron
steel – plastic – aluminium
liners
warships
cargo ships
MATERIALS ON THE MOVE
MATERIALS IN SPACE
alloys
aluminium
new materials
Constructing and investigating
TRANSPORTING MATERIALS
crude oil
tankers pollution
MATERIALS AND LAND TRANSPORT
first cars
chassis
history
bicycles
iron
steel
alloys
composites
history
recycling
composites
modern cars
steel
alloys
plastics

Glossary

Ailerons
Flaps, on the wings of aircraft, used to control flight through the air.

Alloy
A material made from a mixture of two or more metals, or a metal with another substance.

Aluminium
A lightweight white metal. The most common metal in the Earth's crust.

Bond
A term used to describe how minute portions of matter are fixed together.

Cane
Hollow stems of giant grasses and reeds, such as bamboo.

Canvas
A strong cloth made from flax or hemp.

Carbon
A non-metallic element found in the form of diamond and graphite.

Catamaran
A boat with two hulls.

Ceramic
A tough material made from clay and other materials dug from the ground.

Chassis
The rigid framework forming the base of a car.

Composite
A material made of two or more different materials.

Corrosion
Wearing away.

Daggerboard
Board which prevents a boat from being blown sideways by the wind.

Flammable
Easily set on fire.

Gravity
The force that tends to draw all bodies on the Earth's surface or in the Earth's sphere towards the centre of the Earth.

Hull
The main body of a ship.

Leather
Material made from animal skins.

Linen
Cloth woven from flax plants.

Lithium
A metallic element used in alloys.

Magnesium
A light, white metallic element.

Mass
The amount of matter in an object. The units of mass are grams, kilograms and tonnes.

Mine
An explosive device used to destroy ships.

Papyrus
A plant used to make an early form of paper.

Particle
An extremely small portion of matter.

Pneumatic
Filled with air.

Titanium
A dark grey, lightweight metal.

Books to read

The Encyclopaedia of Aviation ed. by P Beaver (Octopus, 1986)

Military Technology by C Smith and B Harbor (Wayland, 1991)

Land Transport by M Dixon (Wayland, 1991)

Ship Technology by M Lambert (Wayland, 1989)

Ships and Submarines by M Grey (Franklin Watts, 1986)

The Car by D Clark (Marshall Cavendish, 1978)

Car Factory by W Pepper (Franklin Watts, 1984)

Pedal Power by P Lafferty and D Jefferies (Franklin Watts, 1990)

The Revolution in Industry by R Kerrod (Gloucester Press, 1990)

Space Shuttles by I Graham (Gloucester Press, 1989)

Places to visit

Welsh Industrial and Maritime Museum
Bate Street
Cardiff

National Motor Museum
Beaulieu
Hampshire

Haynes Sparkford Motor Museum
Sparkford
Nr Yeovil
Somerset
England

Fleet Air Arm Museum
Yeovilton
Somerset

Royal Air Force Museum
Hendon
London NW9

Maritime Museum
Albert Dock
Liverpool

National Maritime Museum
Greenwich
London SE10

The Historic Dockyard
Chatham
Kent

Science Museum
Exhibition Road
South Kensington
London SW7 2DD

HMS *Belfast*
London SE1

SS *Great Britain*
Bristol
Avon

Organizations to contact

USA
Ford of America
300 Schaefer Road
Dearborn
Michigan M148121

Michelin America
Michelin Tyre Corporation
340 Crossways
Parkdrive, Woodberry
New York 11797

AUSTRALIA
Ford Motor Co of Australia
Private Mail Bay 6
Campbellfield
Victoria 3061

Remember to send a stamped addressed envelope with your enquiry.

Index